The TANNERS

An animal hide

COLONIAL AMERICAN CRAFTSMEN

The TANNERS

WRITTEN & ILLUSTRATED BY

Leonard Everett Fisher

DAVID R. GODINE
Publisher • Boston

This is a Godine Storyteller
published in 1986 by
David R. Godine, Publisher, Inc.
Horticultural Hall, 300 Massachusetts Avenue
Boston, Massachusetts 02115

First published in 1966 by Franklin Watts

ISBN: 0-87923-609-4
LC: AC66-10136

First printing
Printed in the United States of America

Colonial American Craftsmen Series

THE DOCTORS

THE PAPERMAKERS

THE SCHOOLMASTERS

THE TANNERS

In Preparation

THE CABINETMAKERS

THE PRINTERS

THE SHIPBUILDERS

THE SILVERSMITHS

A Short History

The *HISTORY*

TANNING IS THE PROCESS BY WHICH animal hides are made into leather. The craft of tanning goes back thousands of years.

Long before the beginning of history, primitive peoples roamed the great forests, plains, and tundras of the world, searching for food. The men and boys were hunters. These peoples' whole lives depended on successful hunting. Nearly all of a slain animal was used by early man. The meat served for food. Many of the bones were shaped into tools and weapons. The skins, dried and stiff, were worn draped around people's bodies as protection against wind, snow, rain, and the sun's heat. But these skins must have been uncomfortable in their stiffness, and early man may have learned to rub animal fat into them, to soften them.

Leather came later, but still early in the history of civilization. Some of the earliest historic peoples, the Sumerians, made leather. An Egyptian tomb painting of fifteen hundred years before Christ shows men working with leather, and an ancient Babylonian script gives a recipe for tanning the skin of a young goat. The milk from

LEF

An Indian buffing a tanned skin with a bone tool.

a yellow goat, oil, alum diluted in grape juice, and the acid from oak bark were among the ingredients used in this recipe.

The *HISTORY*

So it was not a particularly historic occasion when Experience Miller, a tanner, arrived in England's American colony of Plymouth in 1623. Experience Miller may have been the first member of his craft to come from civilized England to the uncivilized New World, but he was not the first person in the American wilderness to make skins into some form of leather. The Indians had various ways of treating deer hides for clothing. One way was to wash and scrape the hides, bury them until the hair was loosened, then sprinkle them with powdered rotten wood to tan them.

The Indians knew a great deal about making leather for moccasins, leggings, and cloaks, but they knew nothing about the kinds of leather needed for shoes, saddles, harnesses, gunpowder pouches, gloves, and the many other things the early American colonists needed and were accustomed to having. Leather goods and many pieces of leather were brought to the new land by the colonists themselves. In time, more leather pieces

LEF

were ordered from England, but it took so long for the shipments to arrive — if they came at all — that few people could afford to wait for them. Since there were no skilled tanners among the colonists before 1623 — and only one more, up to 1630 — and since leather shipments from England were undependable and the Indians' work was unsatisfactory, many of the settlers made whatever leather they needed.

Much of this homemade leather was of poor quality, although many a colonial farmer continued to make it for the next one hundred years. Nevertheless, no one knew better than these early colonists, whose first order of business was to stay alive, that anything poorly fashioned for their use could mean disaster.

Fine, strong leather could be made from the skins of elk and deer and many other wild animals of the American wilderness, and there were enough of them to supply an army of tanners. Still, all over the colonies there was a need for well-made leather.

In 1630, ten years after the Pilgrims landed at Plymouth, English Puritans founded the village of Boston in the Massachusetts Bay area of New

LEF

Peter Minuit trading with the Indians.

England. In that same year Francis Ingalls, a tanner from Lincolnshire, England, established a leathermaking factory, or *tannery,* at Lynn, a tiny settlement northeast of Boston.

During this same period of English settlement, 1623 to 1630, the Dutch West India Company was busy organizing and colonizing a large area of the forest some two hundred miles southwest of Boston. The Company's chief business was exporting beaver skins. The heart of this activity was the village and fort of New Amsterdam, founded on the tip of Manhattan Island in 1626.

Peter Minuit was the man most concerned with the Dutch fur trade. He was the governor and director general of the entire colony of New Netherland, which included New Amsterdam, the capital. A resourceful man, he bought the whole island of Manhattan from the Indians for a handful of trinkets and a bolt of cloth, worth twenty-four dollars. But his resourcefulness did not end there. He later turned his attention to improving his colony's tanning methods.

Shortly before 1633, when he was replaced by Governor Wouter Van Twiller, Peter Minuit built the first machine used in America for the

LEF

Reconstructed Minuit oak-bark-grinding machine.

tanning process. It was a horse-driven stone mill for grinding oak-tree bark, an ingredient in tanning. Soon various small tanneries sprang up in and near New Amsterdam.

In 1664 the Dutch, under Peter Stuyvesant, surrendered their colony to the English, and both New Netherland and New Amsterdam were renamed New York.

Within five years the inhabitants of the little city of New York began to complain about the awful smells that came from the numerous tanneries. Mayor Steenwyck decided that the city's leathermakers should all be located in one place far away from the homes and shops of New Yorkers. Accordingly, an area called the Swamp, situated near what is now the Manhattan side of Brooklyn Bridge, and just south of present-day City Hall, was set aside for this purpose.

For some 275 years the Swamp was one of the important tanning centers of America. But tanning became an ever growing craft in other parts of America, too. From the small, desperate beginnings at Plymouth, Lynn, and New Amsterdam it has today become a giant industry in America.

Crossbeam
support
Rotating
post
Grinding
stone
Axle
Channel filled with oak bark
LFF

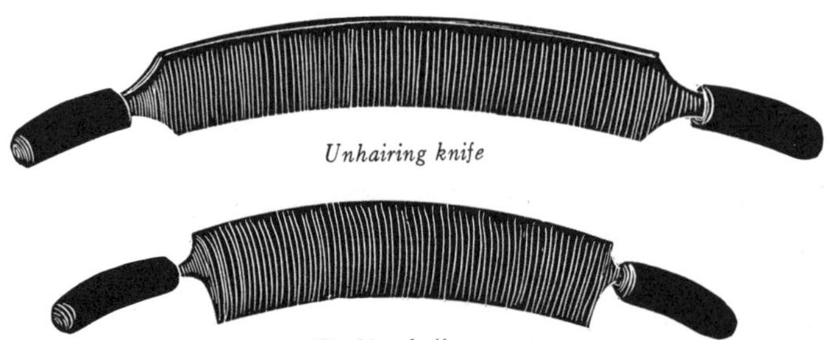

Unhairing knife

Fleshing knife

How the Tanners Worked

Buffer

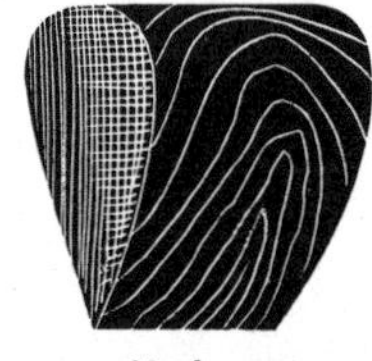

Sleeker

Sleeker

Sleeker

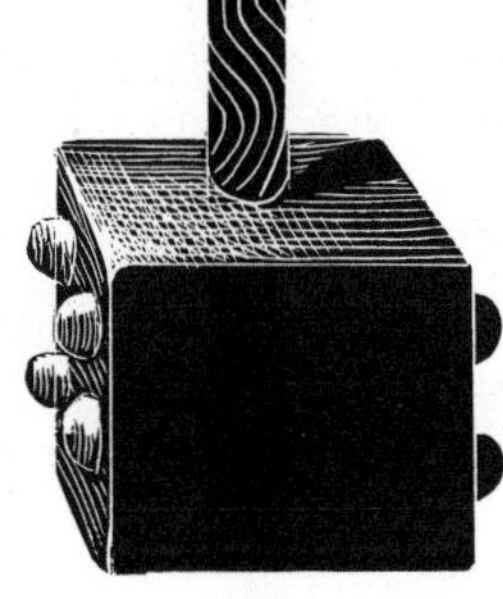

Mace

Vat hook

The *TECHNIQUE*

THE TANNING OF ANIMAL SKINS in Colonial America, as elsewhere in the world, was a long, hard, messy, smelly job. Not all tanners worked alike or produced the same type and quality of leather. There were always variations.

Usually skins were brought to the tanner not long after they had been stripped from the dead animal. Sometimes the hides had been dried. Sometimes they had been salted to keep them from rotting. At other times they were so fresh that they were delivered unsalted. In any event, a good tanner preferred to do his own preserving, or *curing*. Many days and weeks of preparation were needed before a hide could be tanned.

Upon receiving a hide, the first thing the tanner did was to mark it near the tail end with the initials of its owner, if it was to be returned later as leather. In that way the tanner knew where it was to go when it was ready.

Next, the raw and grimy hide, with its hair still on, had to be cleaned, and any salt it held must be rinsed away. Working on the bank of

A typical small colonial tanyard.

Tanning
pits

A tannery worker with hides.

a stream, the tanner washed and soaked, and soaked and washed, the hide. When it had been thoroughly cleaned and softened, it was ready to be cured permanently.

The *TECHNIQUE*

The most widely used method of curing a hide was called *liming*. The lime was a sticky solution of calcium oxide and water. This calcium oxide was made by heating limestone and seashells in very hot ovens, or kilns. Hides were placed in a pit that contained an already used solution of lime. There they were left to soak for two or three days, and during this time were stirred often. The lime solution contained bacteria from hides that had previously been soaked in it, and these bacteria caused the hair roots of the immersed hides to loosen.

After several days of soaking in the lime solution the wet hides were removed and stacked in a pile to drain. Then they were returned to the liming pit. This process went on for several weeks or until such time as the loosened hairs could be easily scraped off without injuring the skin.

Some tanners loosened the hair on heavy ox-

LEF

Washing a skin in a stream.

hides by piling them, wet, in heaps for a few days, then hanging them over poles in a smoke-house where low fires were kept burning. This treatment made the skins sweat, and rot ever so slightly, and the hairs began to fall out.

Whichever method was used, when the tanner was able to rub off some hairs easily with his bare hand he knew it was time to take the next step — cleaning all the hairs from the hides. Each hide was stretched over a tilted, half-rounded, narrow table called a *beam*. Using a blunt tool, the tanner pressed the water out of the hide. Next he picked up an *unhairing knife*, which was about two feet long, with two handles and a sharp, curved lower edge. Moving this knife back and forth, he scraped the loosened hairs off, then rinsed this unhaired, or *grain*, side of the hide with a bucketful of water.

Later the hairs were collected from the base of the beam, where they had fallen, and were sold to an upholsterer or anyone who wanted to stuff pillows. Or they were bought by a plasterer, who mixed them with his mortar to hold it together.

LEF

Unhairing a hide.

Now the tanner flipped the hide over to the other side and cleaned it of any remaining flesh with a *fleshing knife.* This had two edges — a concave edge for scraping, and a convex edge for trimming off parts of the hide that could not be easily tanned.

Next the hide was put for some time into another soaking solution called a *bate.* This solution removed the lime that had worked into the skin, and made it even softer. At last, after a final scrubbing and rinsing, it was cured and ready for the *tanning pit.*

The word "tanning" comes from the same root words as *tannin,* or *tannic acid.* Tannin is the chemical substance that prevents a cured, clean animal hide from further rotting and turns it into leather — of finer texture than hide, and much more durable. Tannin can be extracted as a yellowish powder from the barks of certain trees —oak, chestnut, hemlock, sumac, spruce, and some others. These barks are called *tanbarks.* Although a variety of tanbark trees grew in great abundance in the American forests, the chief tannin-containing tree known to the colonial tanners was the oak.

Beam
LEF

Fleshing a hide.

The *TECHNIQUE*

A hide was not thoroughly tanned until it had been completely saturated with tannin. It had to soak for many months in a pit filled with a tanbark solution. The colonial tanners' pits were huge hogsheads or wooden boxes, six feet long and several feet deep, sunk into the earth near a stream. They were usually arranged in rows with narrow paths between them, along which the tanner walked to attend to the soaking hides.

After the hides had been cured, the tanner filled a pit with a mixture of ground-up oak bark and water, called *ooze.* Into the ooze several cured hides were dumped. Over the weeks they were stirred, or *handled,* from time to time and were occasionally lifted out to be drained. This work was done with long-handled, blunt-pointed poles called *vat hooks.* At this point more tanbark was added to the ooze in order to strengthen the solution. Then the hides were pushed back into the pit. The constant stirring, draining, and adding of tanbark went on for many weary months. It was a hard, muscle-stretching business for the tanner, for the hides were water-soaked and almost backbreakingly heavy to move.

Turning hides in the pit.

The *TECHNIQUE*

To be sure that a hide had fully turned to leather, a tanner snipped a piece from it. The part that had absorbed the tannin was a deep brown. The untanned part was still white. Not until a hide was brown all the way through could it finally be removed from the pit. Then it was drained, stretched on a curved piece of wood, or *horse,* scoured and beaten smooth, and allowed to dry.

If thick hides were being prepared for shoe sole leather, they were taken out of the ooze when they were half tanned and were put into another pit lined on the bottom with used tanbark. On top of this a layer of fresh tanbark was spread, then alternate layers of hides and tanbark were added until the pit was filled to within two feet of the top. The tanner then covered all of this with a one-foot layer of old tanbark and jumped up and down on it until the layers were pressed tightly together. For the next three months small amounts of water were added to the layers of tanbark and hide, to keep them moist and allow the tannic acid from the bark to do its work. By the end of this time the hides had turned to leather and were taken from the pit to dry.

Vat hook
Tanning pit
LEF

Packing down tanbark and hides in a pit.

The *TECHNIQUE*

Unless it was to be used for shoe soles, the stiff, dry leather from the tanning pits had to be softened, cleaned, and otherwise prepared for the leatherworking craftsmen who were to use it to make a variety of articles. The craftsman who did the final finishing of the leather was called a *currier.* His work was called *currying.* In most instances in Colonial America the currier and the tanner were the same person.

Like the other processes in tanning, currying was done in a variety of ways. Usually the first step was soaking the leather in water until it was soft. Then it was shaved with a currying knife until it was of equal thickness over its whole surface. Next it was placed on a slanting table and scoured with a brush and water on both sides, then rubbed with the edge of a smooth stone set in a handle. The water was then pressed out of the leather with an iron *stretcher.*

Now the currier rubbed into the leather a mixture of tanner's oil and tallow, a grease made from animal fat. He then trampled the leather on a *hurdle,* to soften it further. The hurdle was a screen about a yard square, made of basket

LEF

A currier tramples
the leather on a hurdle.

twigs or narrow wooden strips woven about three inches apart.

Next, the currier scraped the leather clean. If it still was not soft enough, he oiled and greased it again, trampled it, cleaned it, and mopped up any excess oil and grease with sawdust. Then, for good measure, he pounded it with a four-sided hammer with knobbed surfaces, called a *mace.* Sometimes the currier continued to oil the leather and buff it with pumice, a *sleeker* — or smoother — and a brush, not only to make it soft and supple, but also to bring out its natural grain and color and give it a handsome luster.

Often the leather was dyed one of several different colors. This too was a hard and messy job in Colonial days. It was done by a craftsman called a *dyer,* after the hide had been tanned and before it was curried.

When all the necessary finishing had been completed, the leather was ready to be returned to the original owner of the hide or, if it was the tanner's property, to be sold to any of the many leatherworkers who carried on their craft all over the Colonies.

LEF

Pounding the leather with a mace.

The *TECHNIQUE*

The entire process of colonial tanning was done by hand. The only exception was the horse-driven oak-bark-grinding mill first used in America by Peter Minuit, and later adopted by other colonial tanners who made their own variations of it.

Tanning was a long-drawn-out, dirty business that prospered and grew during a period in American life that was not noted for good times and easy living. Yet the early American tanners were so skillful that some of their products played an important part in making a free nation of the American colonies.

Among these products were sheets of very fine parchment, or *vellum*.

To be sure, ordinary parchment was made by a *parchmentmaker*, from the untanned skins of sheep, lambs, and goats. Vellum, on the other hand, was usually made from the untanned skins of very young calves.

Nevertheless, it was the colonial tanner, whose chief business it was to make tough leather for shoes, boots, saddles, pouches, and similar useful, everyday objects, who also prepared the fine skins for the parchmentmaker.

Mace
LEF

A parchmentmaker rubbing a cured skin with pumice. In the background a cured skin is stretched in preparation for shaving down to paper-thin parchment

The parchmentmaker, in turn, painstakingly made vellum, with its superior, smoothly finished writing surface. On such a finely crafted colonial product were written the two great documents of American liberty and government — *The Declaration of Independence* and the *Constitution of the United States.*

LEF

Tanners' Terms

BATE — A solution in which a hide was placed after liming, to soften the skin and remove the lime.

BEAM — A tilted, half-rounded table on which a hide was placed for unhairing and fleshing.

CURING — The process by which an animal hide was preserved and made ready for tanning.

CURRIER — The craftsman who put the final finish on the tanned leather, softening and smoothing it.

CURRYING — The process of finishing the newly tanned leather, to make it soft and pliable.

FLESHING — The process of cleaning scraps of flesh from a hide with a curved knife.

GRAIN SIDE — The unhaired side of a hide.

HANDLING — Stirring the hides as they lay soaking in the tanning pits.

HORSE — A curved piece of wood on which a wet, newly tanned hide was stretched.

HURDLE — A woven screen on which leather was trampled, to soften it.

LIMING — Soaking a hide in a lime solution, to loosen the animal hair as a first step to curing.

OOZE — A solution of ground-up tanbark and water, in which hides were soaked until they were tanned.

SLEEKER — A tool for smoothing leather, used in finishing the tanned hide.

TANBARKS — The barks of certain trees — oak, chestnut, hemlock, sumac, spruce, among others — which contain tannin.

TANNIN — A chemical substance extracted from tanbark that turns a hide into leather.

TANNING PIT — The pit holding a solution of tanbark and water, in which the hide was soaked until it was thoroughly tanned, or turned into leather.

UNHAIRING — The process of removing loosened hair from a hide with a curved knife.

VAT HOOKS — Long-handled, blunt-pointed hooked poles for stirring the hides in the tanning pits.

Index

The art of Leonard Everett Fisher bridges both the visual and literary worlds. In a career spanning some thirty-five years he has gained national reputation as a painter, illustrator, author, and educator. His paintings and illustrations have been seen over the years in numerous exhibitions; his books—about 225 titles with which he is connected either as artist, author, or both—are distributed worldwide. In 1973, the New Britain Museum of American Art mounted a major retrospective of his art. His works are represented in the Library of Congress, Smithsonian Institution, Butler Art Institute, Museum of American Illustration, Mt. Holyoke College, and the Universities of Oregon, Minnesota, and Southern Mississippi, among others. A graduate and Winchester Fellow of Yale University, he holds a Pulitzer Prize for painting; the National Jewish Book Award for Children's Literature, the Christopher Medal for Illustration, the Premio Grafico Fiera di Bologna, the Medallion of the University of Southern Mississippi, and numerous citations including those of the American Library Association, the National Council of Social Studies, the American Institute of Graphic Arts, and the Connecticut League of Historical Societies. He has designed a number of United States postage stamps, was a Delegate-at-large to the White House Conference on Library and Information Services, and served overseas with the United States Army Corps of Engineers during World War II. He is Dean Emeritus of Paier College of Art and visiting professor at that institution.

The text of this book has been set on the Linotype in Caslon 137. This face is derived from the great old style cut by William Caslon of London in the early eighteenth century. Caslon types were used widely by American printers during the colonial period and have been called "the finest vehicle for the conveyance of English speech that the art of the punchcutter has yet devised."